U0269019

作者自画像
私人收藏
2015

Wang Yun   Paintings Originated from Settlement Plans

王 昀　聚落平面中的绘画

中国电力出版社
CHINA ELECTRIC POWER PRESS

目 录

Contents

空间的平面性、身体性与兼具空间指向性的平面绘画

Planarity and Corporality of Space and Plane Drawing with Space Directionality

空间的平面性与平面绘画的空间性

呈现为二维平面状态的聚落平面图，是聚落中所有居民"空间概念""集合"的同时，也是生活在聚落中那些居民们自身整体"经验"的诗学呈现。尽管从表面上看，聚落平面图本身所呈现的状态是一种二维的平面状态，究其内容，也是以"点、线、面"为基本形式所构成的"图案"，或许这些图案还是一个"美丽"的图案。但拥有建筑学含义的聚落平面图中所呈现出的这些"图案"，其实与通常平面绘画中所意味的"画面"与"图案"不同。这些拥有建筑学意义的平面图所呈现的以"点、线、面"为基本形式所构成的"图案"，究其本质，乃是拥有三维深度、四维指向性的空间表述与形态呈现。

在对拥有建筑学意义的二维平面图进行观测时，二维平面图本身具有让人在意识中完成从平面世界到空间世界过渡的功能，同时还能够让人从二维平面图中从那些由"点、线、面"所构成的聚落平面图的"画面"与"图案"中，获得一种潜在的空间感受。

我们在这里所提到的空间、平面，是一种针对"自己"（或处于镜像状态的他者）的经验所进行的描述。

拥有建筑学意义的平面图中呈现的"点、线、面"本身，是作用在头脑中的划痕，也是在意识中针对空间关系的表述所采用的记号，这些记号本身对应着现实中"物"的位置，对应着与现实中"物"与"物"之间的相互关系，表达着意识中对于"空间"的想象表述以及意识中所谓"空间"的构成与组合。

一旦我们如此这样地去思考聚落的平面图，瞬间，我们会将自己的身体直接投射到二维平面中，投射到那些由"点、线、面"图像关系所产生的"空余"（或"余白"）中，并会瞬间地体会到自身的身体正频繁于自身的意识中，与那些制造着"空余"（或"余白"）的"点、线、面"图像本身产生碰撞，进而也就完成了拥有建筑学意义的二维平面图本身所具有的，能够唤起

"空间感"的含义与功能。

如果我们站在这个层面上去理解平面图，并由此扩展到仅仅关注聚落平面图纸的表象，那么，聚落平面图中所呈现的"点、线、面"关系，在表述其自身整体拥有能够与现实空间世界相互对应和相互指向的功能关系的同时，还可进一步启发平面图本身所拥有的在绘画层面上的含义和内容。正是因为通过这个平面图本身与空间之间的转化关系，从而进一步提示于我们：其实可以从单纯的平面绘画中去捕捉一个充满建筑含义的空间世界。

沿着这样的逆向逻辑推演，并进而反复思考：这种兼具并拥有空间指向性，呈现为由"点、线、面"所构成的二维平面的图案表象，不恰恰间接地证明了：绘画本身，不，是抽象绘画本身，只有并必需要站在空间的角度，以空间的视角去构思和理解，才能够真正使抽象绘画本身产生意义。进一步而言，上述的一切思考，也恰恰造就了拥有空间含义的聚落平面图本身，其可以并能够，天然地成为兼具平面和空间意义绘画的开始。

当我们将有空间含义的聚落平面图"降维"（即降低维度）并等同于二维绘画时，不难发现：聚落平面图的绘画本身具有一个非常重要的特点，即每个平面图的内容并不是单纯的平面图形，而是一种从顶部向下，并且纵深与顶面重合，纵深空间的下部被上部遮蔽之后的空间状态；而这种遮蔽使每个平面的、纵深的高度拥有了多样性和不确定性，从而自然也就具有了丰富性。同时由于地面的消失，原本极具重力特征的空间体块具有了漂浮感觉。而一旦将聚落的平面图中的空间体块转换为悬浮于空中状态的"点、线、面"的绘画世界，一个广奥宇宙的空间世界将会扑面而来。由此，空间的平面性与平面绘画的空间性也就在这样的瞬间时刻得到最终的完结。

Settlement plans, in two-dimensional form, serve not only as "spatial concept" and "assemblage" for all inhabitants in the community but also a poetic presentation of their own whole "experience" of inhabitants, who live in the community. From its surface, Settlement plans themselves take on a two-dimensional form and the contents are patterns constituted by basic forms such as "dot, line and plane", while these patterns each may be a "beautiful" image . But patterns presented by Settlement plans with architectural sense, in fact, differ from "pictures" and "patterns" contained in conventional plane drawings. These patterns constituted by basic forms such as "dot, line and plane" with architectural sense, in essence, are spatial expression and form display with three-dimensional depth and four-dimensional directionality.

When observed, two-dimensional settlement plans with architectural sense themselves can not only enable people to transit from plane to space in consciousness but also provide people with underlying feelings of space through two-dimensional settlement plans, to be specific, "pictures" and "patterns" of Settlement plans constituted by basic forms such as "dot, line and plane".

Space and plane mentioned here refers to a description regarding experience for "oneself" (or others in mirror status).

As for "dot, line and plane" with architectural sense, they are scratch marks in brains as well as symbols adopted to describe spatial relationship in consciousness. These symbols themselves correspond to not only position of "objects" in reality but also interrelationship between "objects" and "objects" in reality, expressing imaginary description on "space" in consciousness as well as constitution and combination of so-called "space" in consciousness.

Once we think Settlement plans in this way, then instantaneously we will inject our body directly into the two-dimensional settlement plans and into the blankness caused by interrelation between "dot, line and plane", therefore instantaneously feel that our bodies are constantly, in our own consciousness, colliding with graphs of "dot, line and plane" that produce blankness. Consequently, the meaning and function to provoke "sense of space" contained in two-dimensional settlement plans with architectural sense is thus accomplished.

If we try to understand settlement plans from this level and consequently focus only on the surface of Settlement plans, we may find that "dot, line and plane" presented by Settlement plans not only express their functional relationship of mutual corresponding and directing with space reality but also further inspire implication and content possessed by settlement plans themselves on the level of painting. It is through this transformation between settlement plans themselves and space that further reminds us a specific space full of architectural sense can be captured from simple plane drawings.

Following such a reverse logic, we may further think in a reverse way: this space directionality, which is presented as two-dimensional plane patterns constituted by "dot, line and plane", exactly proves in an indirect way that painting itself, or, abstract painting itself, can truly generate meaning only when viewed with space perspective in conception and understanding. Furthermore, the above thoughts exactly construct Settlement plans themselves with space meaning, which may and can naturally give rise to paintings with both plane and space meaning.

When we decrease dimensions on Settlement plans with space meaning and equal them to two-dimension paintings, we find that Settlement plans themselves have one very significant feature, to be more specific, the content of each settlement plan is not simply settlement plan but a space form from top to bottom, with depth and top overlapping and the bottom part of the depth covered by its upper part. This way of covering gives rise to diversity and uncertainty in terms of the depth height of each plane, naturally bringing about variety. Meanwhile, as ground disappears, space blocks that originally have strong gravity feature take on a floating sense. Once we transform space blocks in Settlement plan into a painted world of floating "dot, line and plane", a vast universe space will embrace you. Consequently, the planarity of space and spatiality of settlement plans, at this very moment, reach final conclusion.

兼具空间指向性的抽象
Abstractness with Space Directionality

抽 象 Abstract -90° Mirror *No.001*
来自非洲阿巴拉克聚落
Abalak, Africa
私人收藏
2015

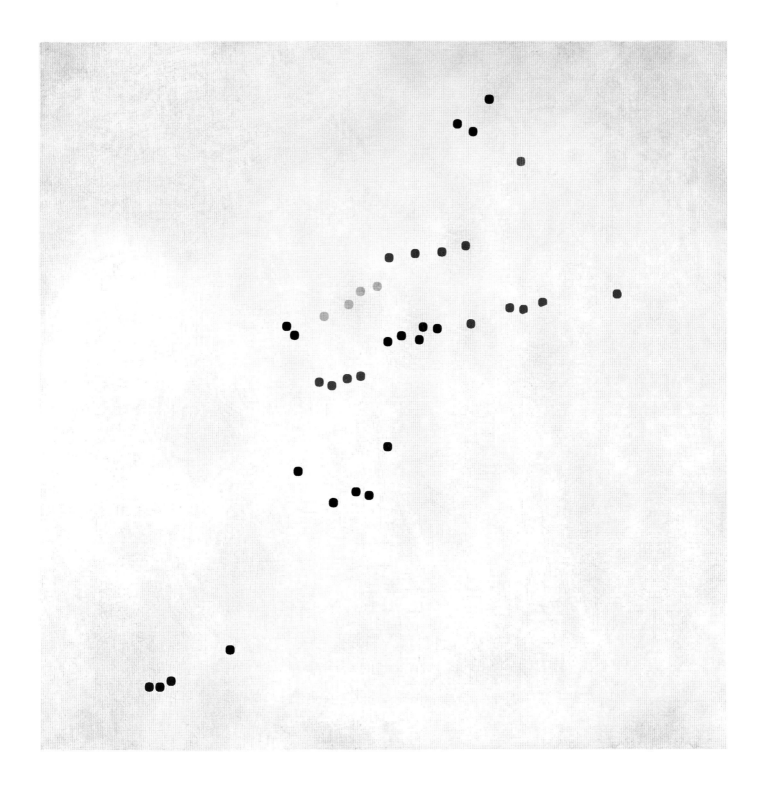

抽 象 Abstract -90° Mirror *No.002*
来自非洲阿泽勒聚落
Azzel, Africa
私人收藏
2015

抽 象 Abstract +90° *No.004*
来自非洲达布努聚落
Akabounou, Africa
私人收藏
2015

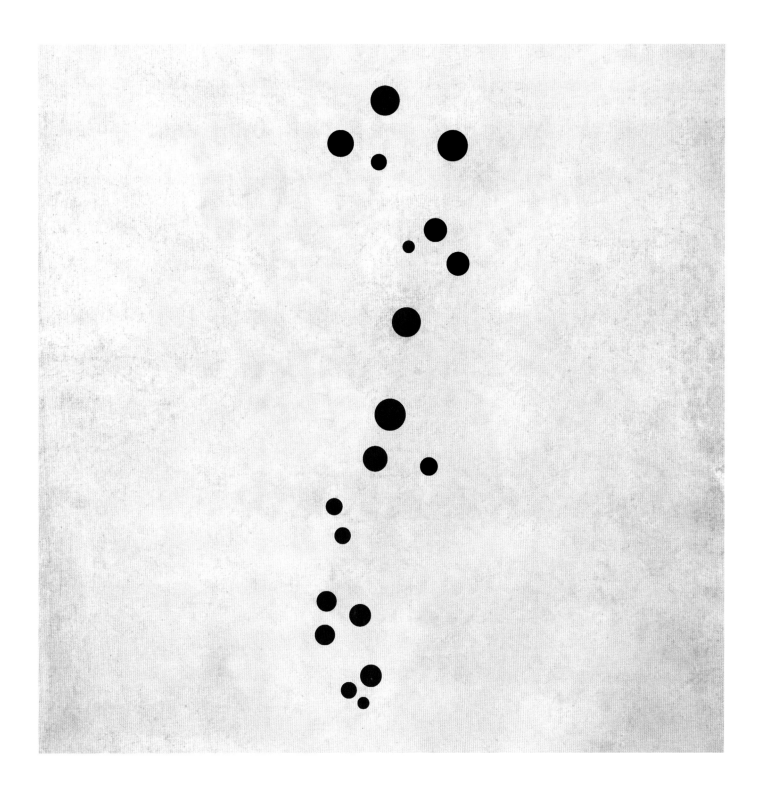

抽 象 Abstract +180° *No.005*
来自非洲博尔博尔聚落
Bolbol, Africa
私人收藏
2015

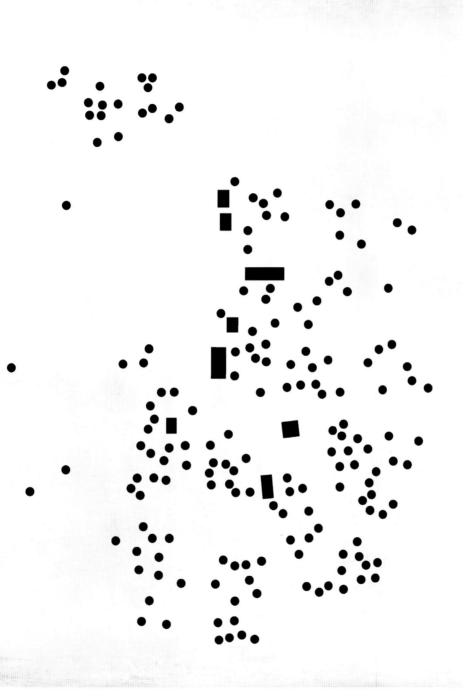

抽象 Abstract No.009
来自非洲鲁贡宾聚落
Rougoubin, Africa
私人收藏
2015

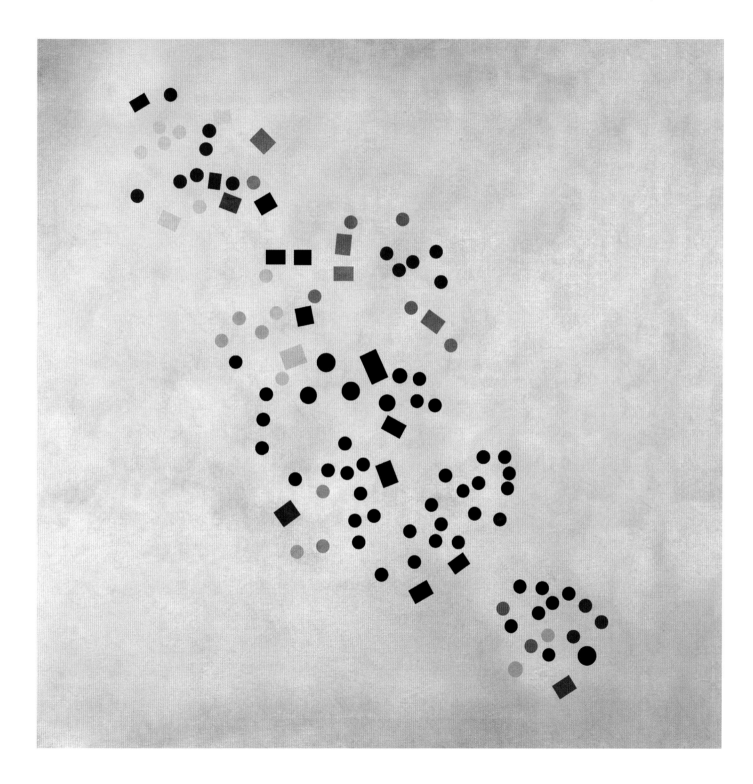

抽 象 Abstract +90° *No.007*
来自非洲坎培玛聚落
Kampema, Africa
私人收藏
2015

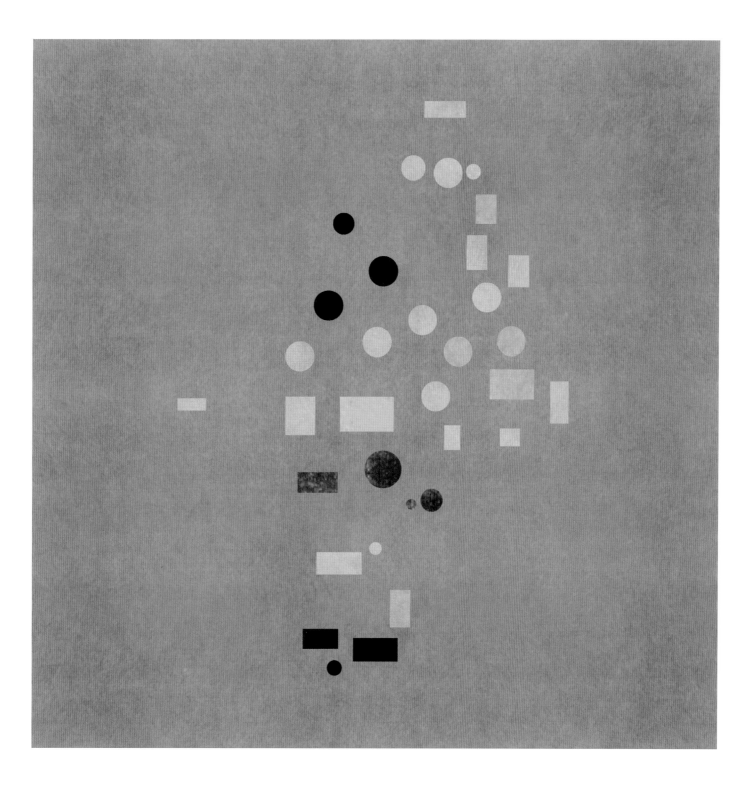

抽 象 Abstract +135° Mirror No.010
来自非洲托斯比克聚落
Toussibik, Africa
私人收藏
2015

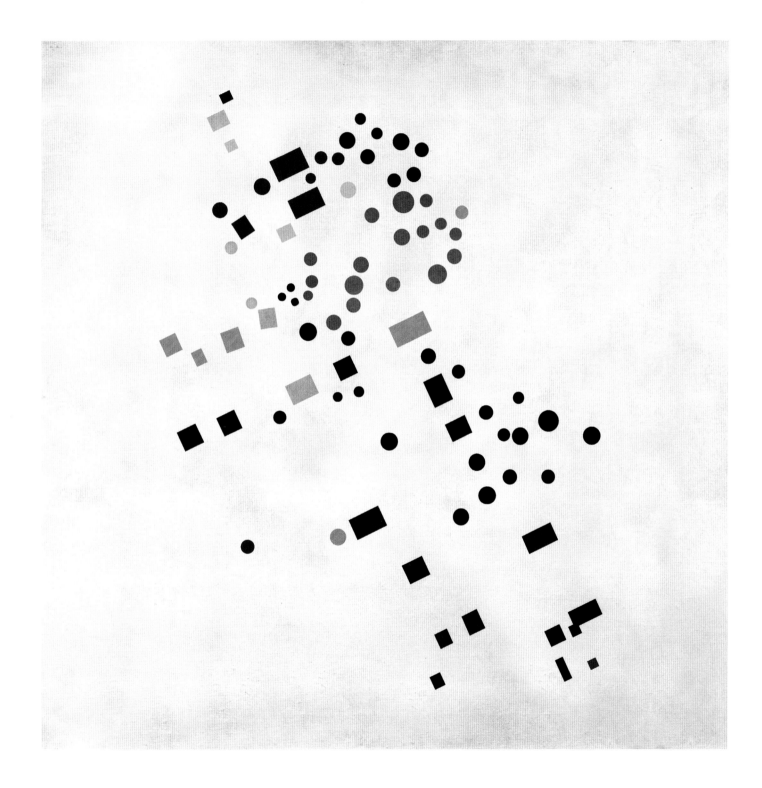

抽 象 Abstract -45° *No.012*
来自中国日月山村聚落
Riyueshan Village, China
私人收藏
2015

抽 象 Abstract No.018
来自中国高走村聚落
Gaozou Village, China
私人收藏
2015

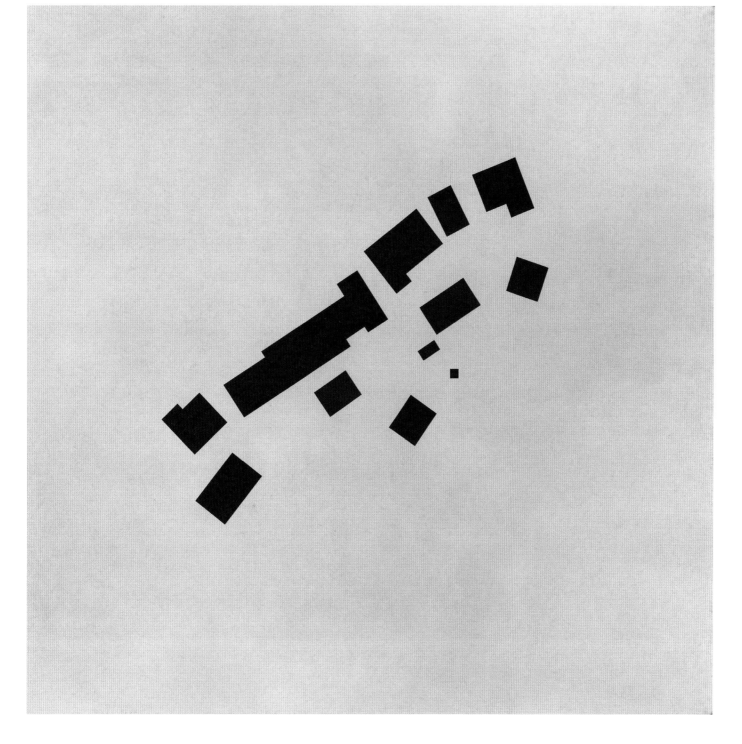

抽 象 Abstract　No.023
来自中国建塘村聚落
Jiantang Village, China
私人收藏
2015

抽 象 Abstract -45° *No.025*
来自中国回库村聚落
Huiku Village, China
私人收藏
2015

抽 象 Abstract　No.027
来自中国漫伞村聚落
Mansan Village, China
私人收藏
2015

抽 象 Abstract　　*No.008*
来自非洲庞博卡聚落
Pomboka, Africa
私人收藏
2015

抽 象 Abstract Mirror *No.028*
来自中国农沙湖聚落
Nongshahu Village, China
私人收藏
2015

抽 象 Abstract -90° *No.032*
来自中国腾梁山聚落
Tengliangshan Village, China
私人收藏
2015

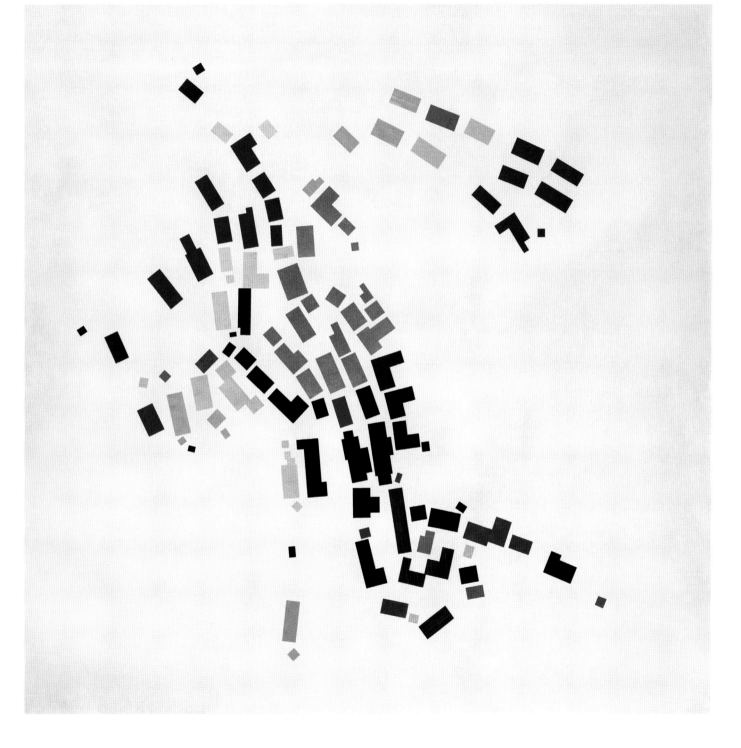

抽 象 Abstract +90° *No.049*
来自印度宁勒塔库鲁姆聚落
Ningle taklum, India
私人收藏
2015

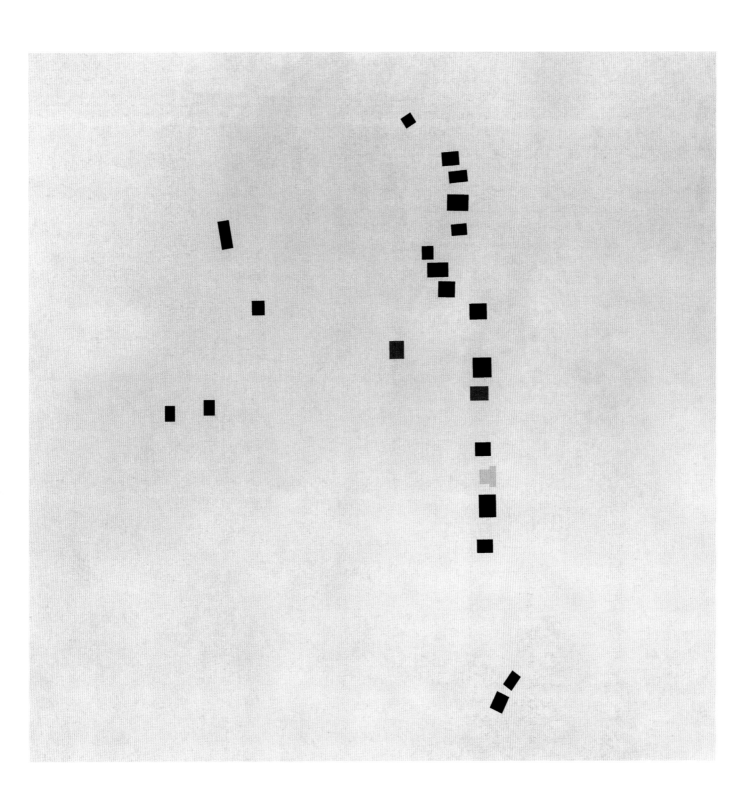

抽 象 Abstract +90° *No.057*
来自印度尼西亚库吉拉图聚落
Kudji ratu, Indonesia
私人收藏
2015

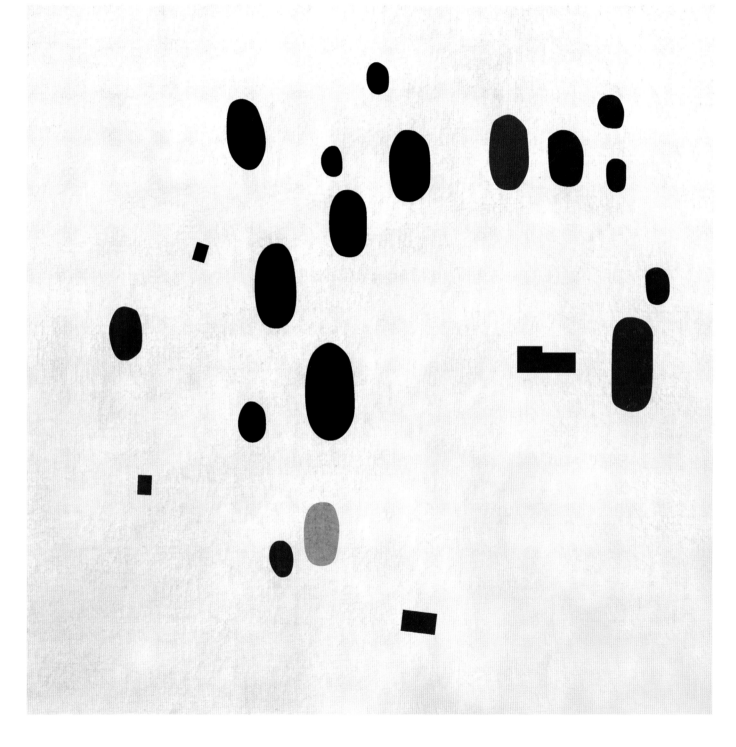

抽 象 Abstract -135° *No.059*
来自印度尼西亚林加村聚落
Lingga, Indonesia
私人收藏
2015

抽 象 Abstract +45° *No.060*
来自印度尼西亚奥尔布布聚落
Oelbubu, Indonesia
私人收藏
2015

抽 象 Abstract -90° *No.062*
来自印度尼西亚杜坎聚落
Dokan, Indonesia
私人收藏
2015

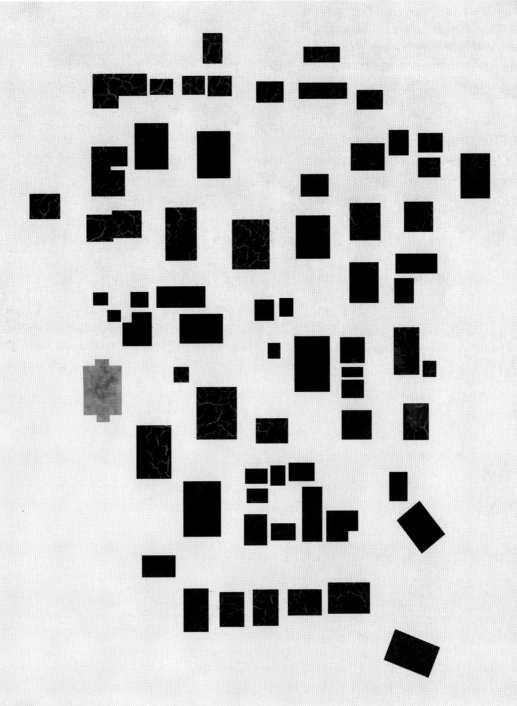

抽 象 Abstract -90° *No.067*
来自印度尼西亚兰博亚聚落
Lamboya, Indonesia
私人收藏
2015

抽 象 Abstract -90° *No.069*
来自印度尼西亚蓝珀聚落
Lempo, Indonesia
私人收藏
2015

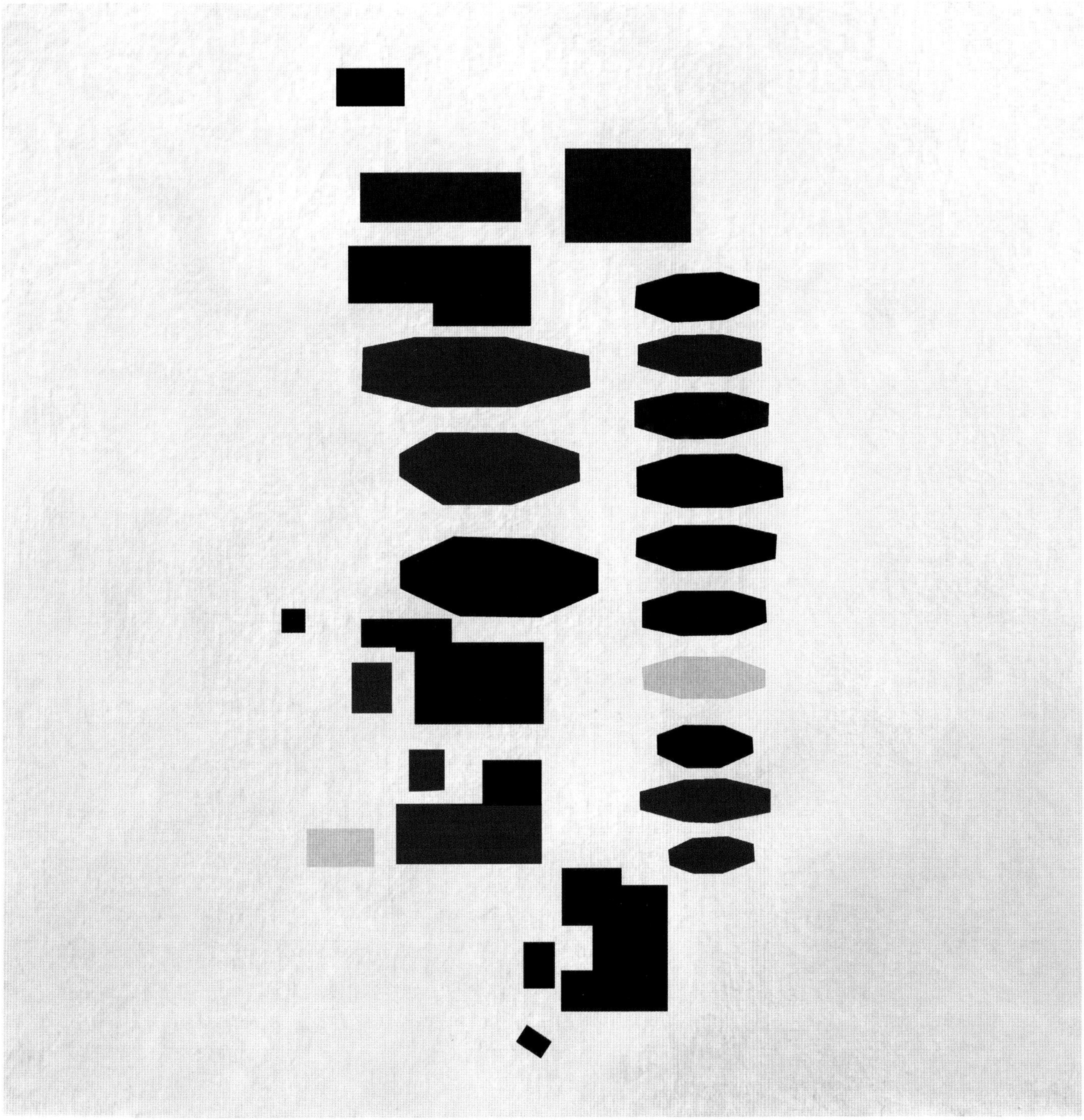

抽 象 Abstract -135° *No.071*
来自印度尼西亚南加拉聚落
Nanggara, Indonesia
私人收藏
2015

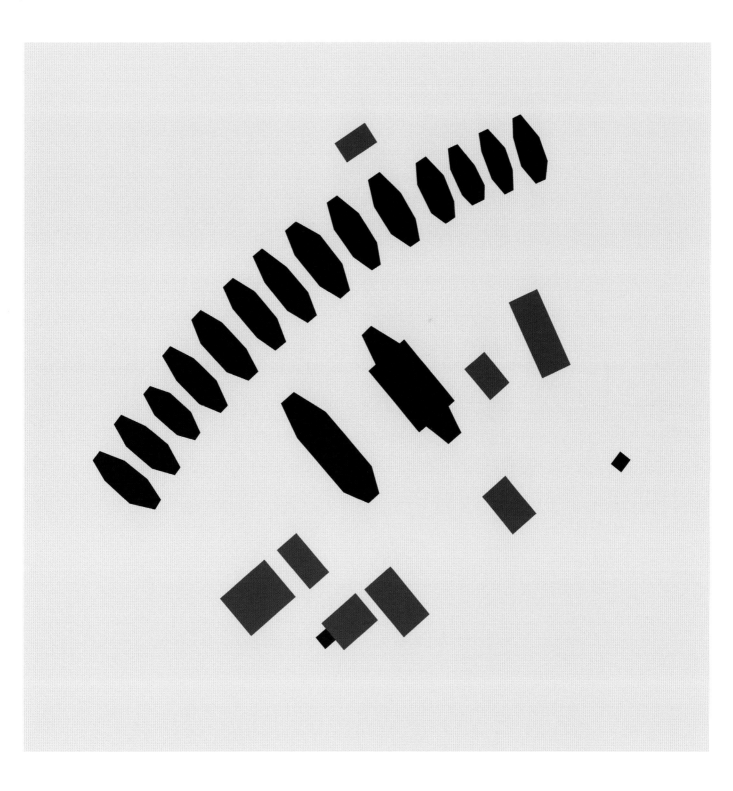

抽 象 Abstract Mirror *No.046*
来自印度马坦瓦里聚落
Matanwari, India
私人收藏
2015

抽 象 Abstract -90° Mirror *No.052*
来自印度西弗利聚落
Shivli, India
私人收藏
2015

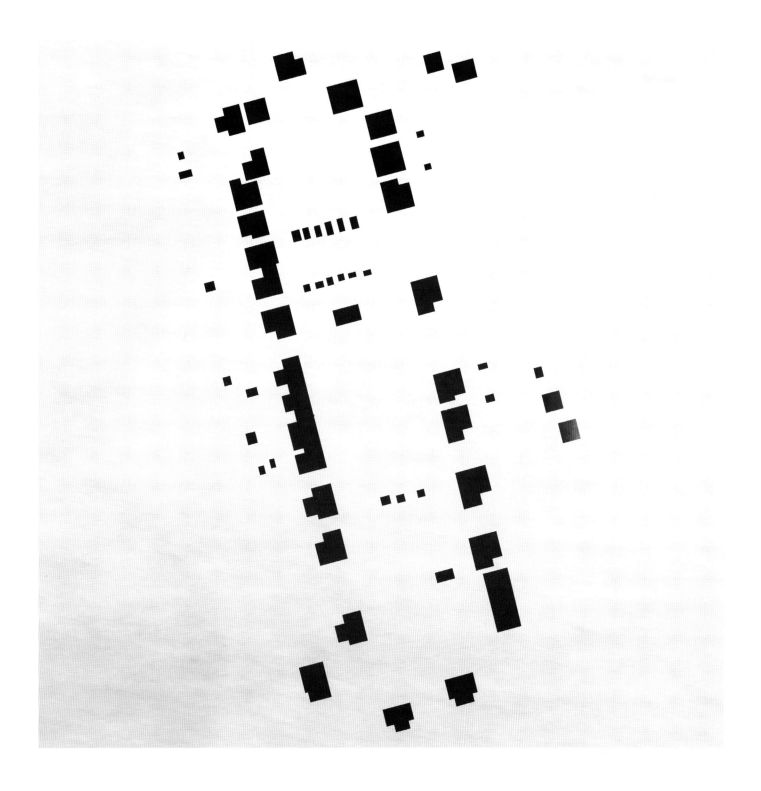

抽 象 Abstract -90° *No.046*
来自印度克里亚特聚落
Keriyat, India
私人收藏
2015

抽 象 Abstract Mirror *No.077*
来自中南美乌尔斯卡斯卡拉聚落
Uros Kaskalla, Middle and South America
私人收藏
2015

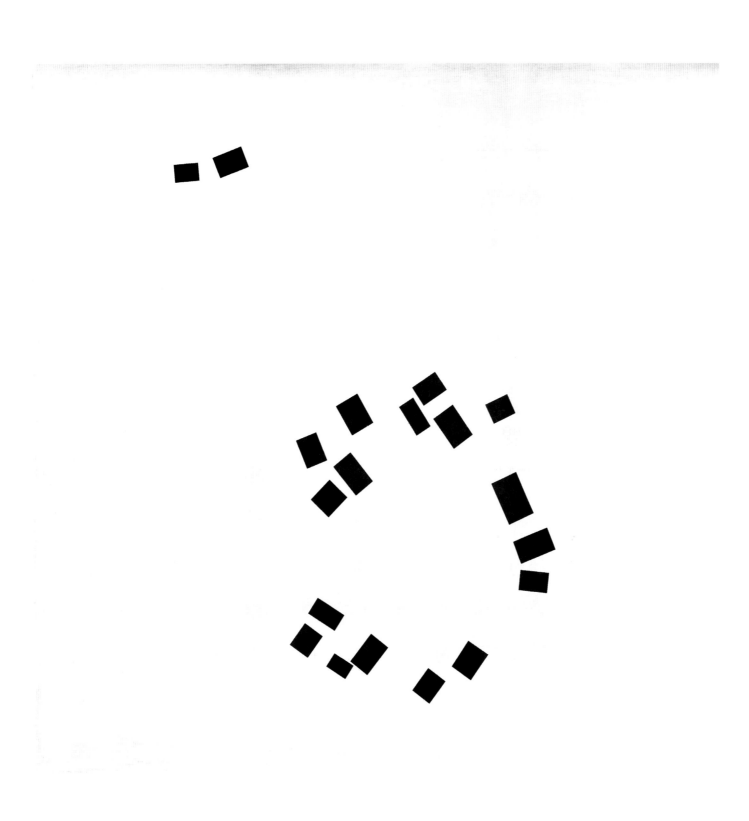

抽 象 Abstract　No.073
来自印度尼西亚萨德/赖比坦聚落
Sade/Rebitan, Indonesia
私人收藏
2015

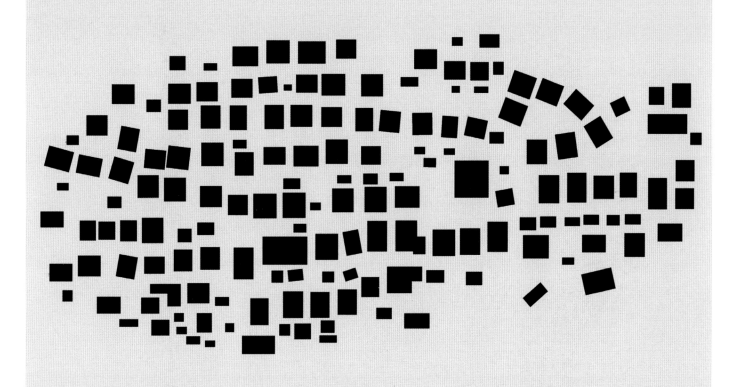

抽象 Abstract　No.042
来自印度坎克沃聚落
Kankewar, India
私人收藏
2015

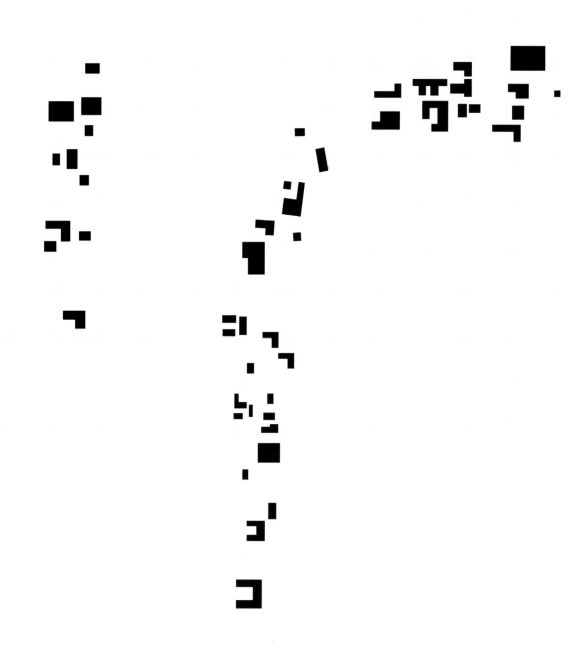

抽 象 Abstract -90° *No.081*
来自中南美圣豪尔赫聚落
San Jorge, Middle and South America
私人收藏
2015

抽 象 Abstract -135° *No.031*
来自中国偏坡村聚落
Pianpo Village, China
私人收藏
2015

抽 象 Abstract   *No.085*
来自巴布亚新几内亚鲁亚聚落
Luya, Papua New Guinea
私人收藏
2015

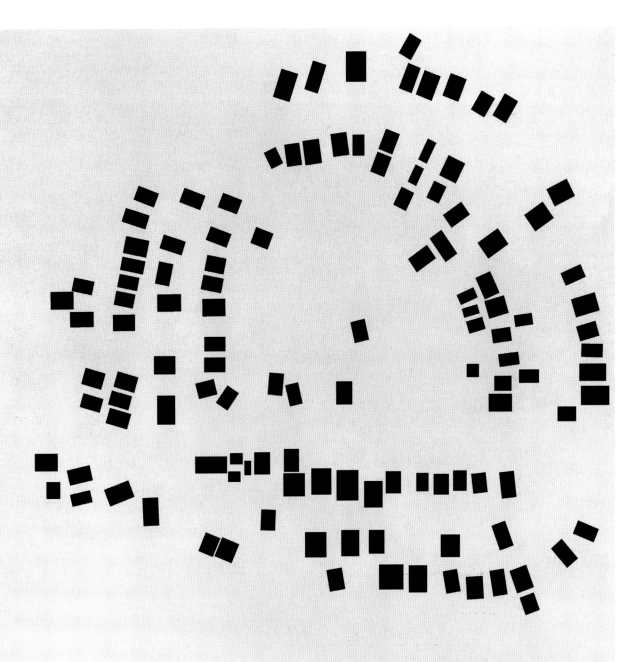

抽 象 Abstract -45° *No.086*
来自巴布亚新几内亚曼杜聚落
Mando, Papua New Guinea
私人收藏
2015

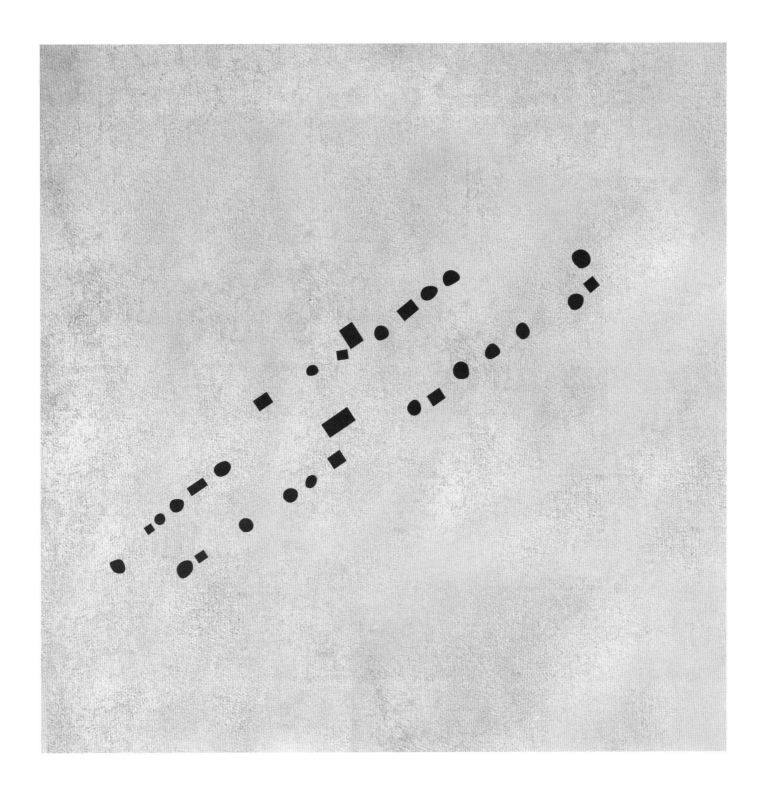

抽 象 Abstract -90° *No.087*
来自巴布亚新几内亚纳帕摩哥纳聚落
Napamogona, Papua New Guinea
私人收藏
2015

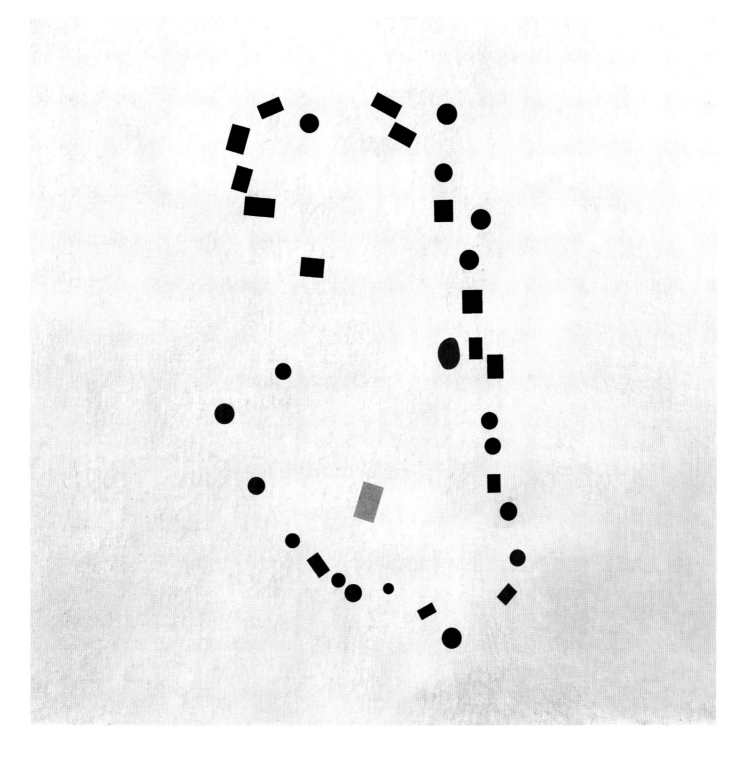

抽 象 Abstract +90° *No.091*
来自巴布亚新几内亚帕兰贝聚落
Palambei, Papua New Guinea
私人收藏
2015

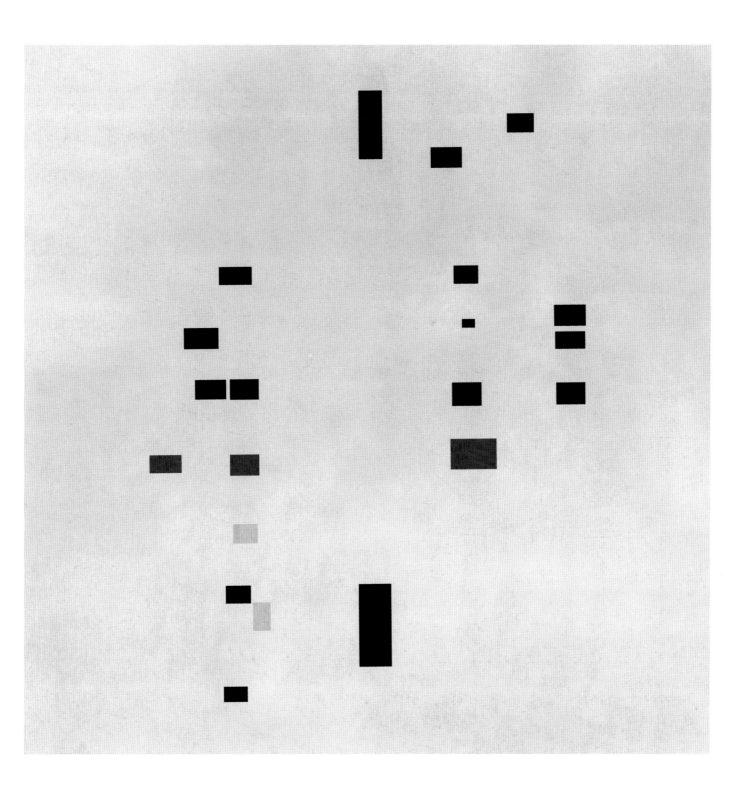

抽 象 Abstract +90° *No.095*
来自中东锡夫里希萨尔聚落
Sivrihisar, Middle East
私人收藏
2015

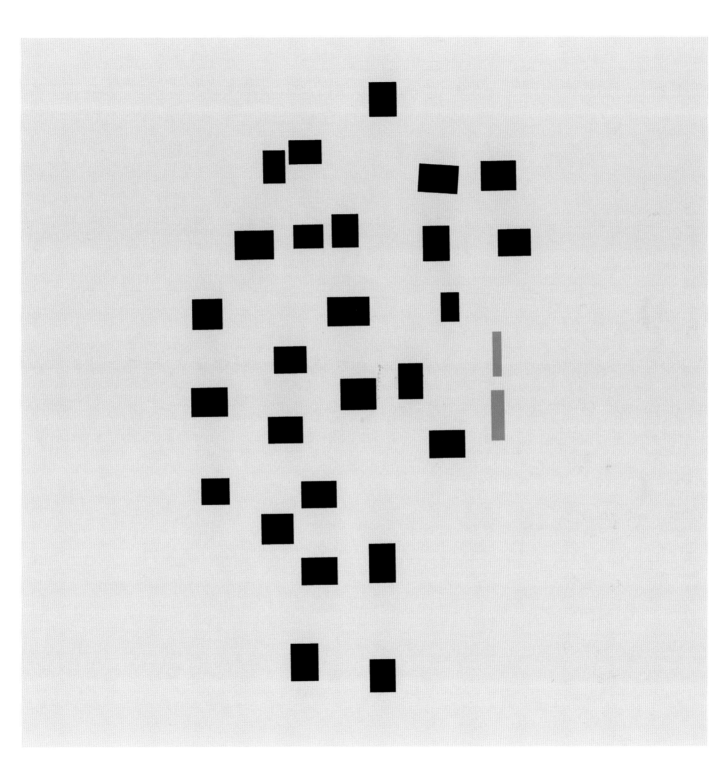

抽取抽象中的抽象
Extract Abstraction from Abstractness

右图是对中国腾梁山聚落总平面图进行"点、线、面"抽象抽取后所获得的聚落住居及建筑物的空间关系图。通过对此聚落空间关系图进行局部再次抽取，获得了"空间的界限 No.515""空间的界限 No.119""空间的界限 No.96""空间的界限 No.121""空间的界限 No.197""空间的界限 No.101""空间的界限 No.138""空间的界限 No.520"等一系列拥有空间关系意味的抽象图。

The diagram on the right is a space relationship graph among community buildings after extracting "dot, line and plane" from the general settlement plan of Tengliang Mountain Community in China. Through a second extraction on the part of this community space relationship plan, we obtained a series of abstractions with space relationship meaning, such as "space boundary No. 515" "space boundary No. 119" "space boundary No. 96" "space boundary No. 121" "space boundary No. 197" "space boundary No. 101" "space boundary No. 138" and "space boundary No. 520".

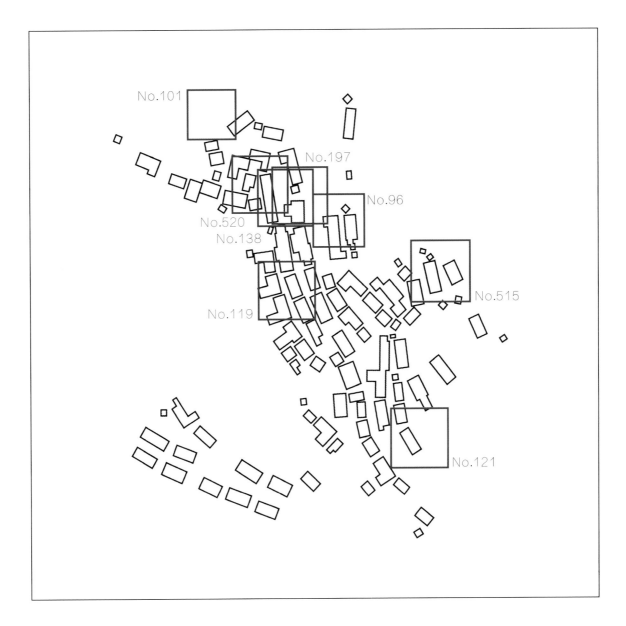

No.101

No.197

No.96

No.520

No.138

No.119

No.515

No.121

抽取抽象中的抽象的位置

Positions to extract abstraction from abstractness

空间的界限 No.515

Space Boundary No.515

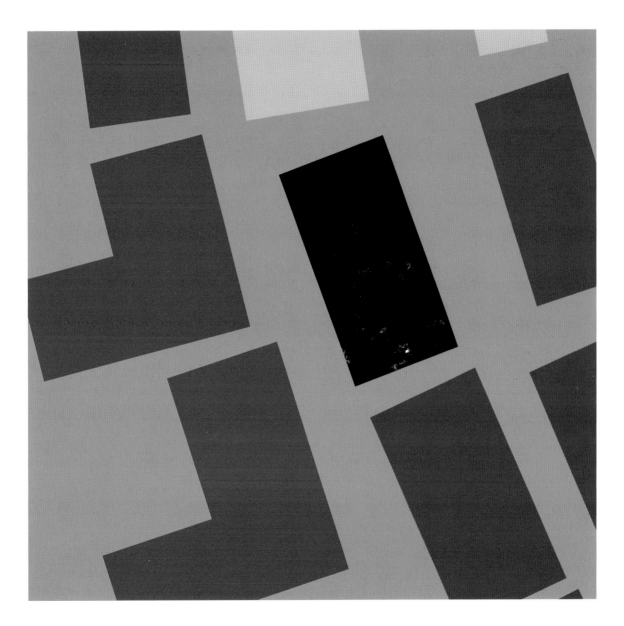

空间的界限 No.119

Space Boundary No. 119

空间的界限 No.96

Space Boundary No. 96

空间的界限 No.121

Space Boundary No. 121

空间的界限 No.197

Space Boundary No. 197

空间的界限 No.101

Space Boundary No. 101

空间的界限 No.138

Space Boundary No. 138

空间的界限 No.520

Space Boundary No. 520

画家介绍

Introduction of Painter

王昀 博士

1985 年毕业于北京建筑工程学院建筑系，获学士学位

1995 年毕业于日本东京大学，获得工学硕士学位

1999 年于日本东京大学获得工学博士学位

2001 年执教于北京大学

2002 年成立方体空间工作室

2013 年创立北京建筑大学建筑设计艺术研究中心

建筑设计竞赛获奖经历：

1993 年日本《新建筑》第 20 回日新工业建筑设计
竞赛获二等奖

1994 年日本《新建筑》第 4 回 S×L 建筑设计竞赛
获一等奖

主要建筑作品：

善美办公楼门厅增建、60 平方米极小城市、石景山财政局培训中心、庐师山庄、百子湾中学校、百子湾幼儿园、杭州西溪湿地艺术村 H 地块会所等

参加展览：

2004 年 6 月参加 "'状态'中国青年建筑师 8 人展"

2004 年首届中国国际建筑艺术双年展参展

2006 年第二届中国国际建筑艺术双年展参展

2009 年参加在比利时布鲁塞尔举办的"'心造'——中国当代建筑前沿展"

2010 年参加威尼斯建筑艺术双年展、德国 karlsruhe Chinese Regional Architectural Creation 建筑展

2011 年参加捷克 prague 中国当代建筑展、意大利罗马"向东方——中国建筑景观"展、中国深圳·香港城市建筑双城双年展等

2012 年第 13 届威尼斯国际建筑艺术双年展中国馆参展

Dr. Wang Yun

Graduated with a Bachelor's degree from the Department of Architecture at the Beijing Institute of Architectural Engineering in 1985.

Received his Master's degree in Engineering Science from Tokyo University in 1995.

Received a Ph.D. from Tokyo University in 1999.

Taught at Peking University since 2001.

Founded the Aterier Fronti (www.fronti.cn) in 2002.

Established Graduate School of Architecture Design and Art of Beijing University of Civil Engineering and Architecture in 2013.

Prize:

Received the second place prize in the "New Architecture" category at Japan's 20th annual International Architectural Design Competition in 1993.

Awarded the first prize in the "New Architecture" category at Japan's 4th S×L International Architectural Design Competition in 1994.

Prominent Works:

ShanMei Office Building Foyer, a Small City of 60 Square Meters, the Shijingshan Bureau of Finance Training Center, Lushi Mountain Villa, Baiziwan Middle School, Baiziwan Kindergarten, and block H of the Hangzhou Xixi Wetland Art Village.

Exhibitions:

The 2004 Chinese National Young Architects 8 Man Exhibition, the First China International Architecture Biennale, the Second China International Architecture Biennale in 2006, the "Heart-Made: Cutting-Edge of Chinese Contemporary Architecture" exhibit in Brussels in 2009, the 2010 Architectural Venice Biennale, the Karlsruhe Chinese Regional Architectural Creation exhibition in Germany, the Chinese Contemporary Architecture Exhibition in Prague in 2011, the "Towards the East: Chinese Landscape Architecture" exhibition in Rome, and the Hong Kong-Shenzhen Twin Cities Urban Planning Biennale. The thirteen Venice International Architecture-Art Biennale in 2012.

www.fronti.cn

图书在版编目（CIP）数据

聚落平面图中的绘画 / 王昀著. -- 北京：中国电力出版社, 2016.2
ISBN 978-7-5123-8575-7

Ⅰ.①聚… Ⅱ.①王… Ⅲ.①建筑艺术－绘画研究 Ⅳ.①TU204

中国版本图书馆CIP数据核字(2015)第279702号
本教材受北京建筑大学设计学学科建设项目资助出版

中国电力出版社出版发行
北京市东城区北京站西街19号 100005
http://www.cepp.sgcc.com.cn
责任编辑：王　倩
封面设计：王　昀
版式设计：张捍平　楚东旭
责任印制：蔺义舟
责任校对：闫秀英
英文翻译：陈伟航
北京盛通印刷股份有限公司印刷 · 各地新华书店经售
2016年2月第1版 · 第1次印刷
787mm×1092mm 1/12 · 9 印张 · 226 千字
印数：1-1500册
定价：78.00元

封面作品：

抽 象 Abstract -45° *No.1915*
来自中国日月山村聚落
Riyueshan Village, China
私人收藏
2015